BACK FROM THE BRINK

# INSECTS

BLACK RABBIT BOOKS | JOANNE MATTERN

# TABLE OF CONTENTS

1

# American Burying Beetle

This colorful beetle lives in the United States. It has an important, but gross, job. The bugs smell a dead animal. They roll the body into a ball. Then they bury the body. The female lays eggs on it. The dead body is food for the beetle's larvae.

The beetle almost died out because houses were built on its land. In 1989, the government protected the beetles. Today, these bugs are coming back to do their dirty jobs.

### Think About It

Why is it important to have insects that clean up dead animals?

# 2

# Hine's Emerald Dragonfly

This dragonfly lives near the Great Lakes. It has big eyes and long wings. It is a great flyer. But by the 1990s, people drained many of the wet places where it lived. It looked like this dragonfly would soon be gone too.

People stepped up to help. The US government made it against the law to kill this dragonfly. Conservation groups went to court. They asked the government to protect the land too. In 2010, they won this legal battle. Now, the dragonfly has a better chance to survive.

**Think About It**

How can people and the government work together to save insects?

3

# Karner Blue Butterfly

This tiny butterfly's larvae only eat one type of flower. That flower is wild lupine. When the flowers began to disappear, so did the butterfly. People knew they had to save the butterfly's habitat.

They planted more flowers. Scientists raised some of the butterflies. Then they let them go into the wild. Slowly, these butterflies began flying around the northeastern United States once again.

Did You Know?
An adult Karner only lives for five days.

4

# Lord Howe Island Stick Insect

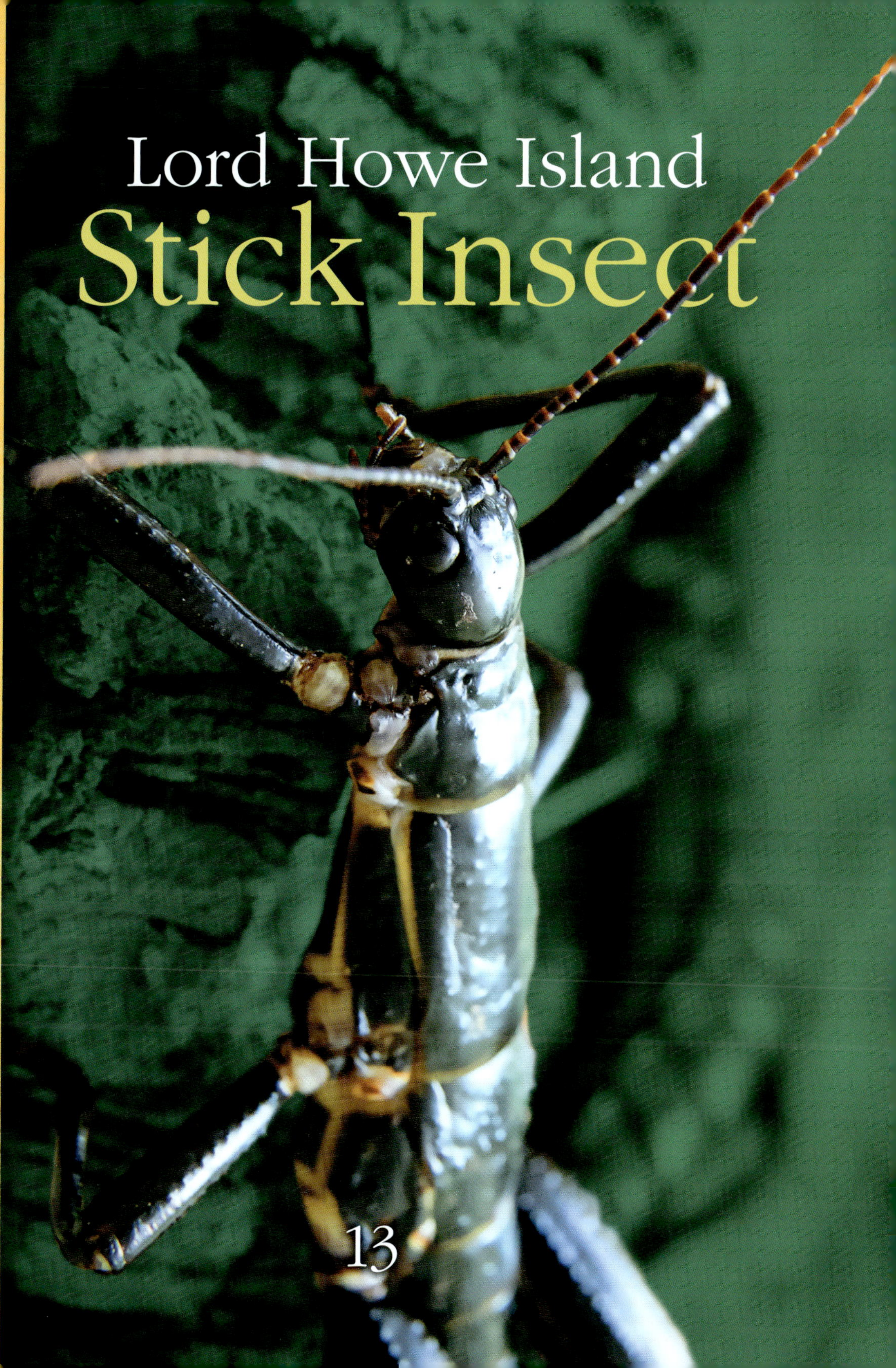

This stick insect only lived on an island near Australia. Then rats came to the island. They ate almost all of the insects. A few stick insects were found in 2001. Australia hurried to save them.

Scientists sailed to the island. They captured a breeding pair of insects. They brought them to a zoo. The insects laid eggs there. Other scientists are getting rid of the rats on the island. Then the stick insect can return home.

Did You Know?
This insect looks like a lobster. Its nickname is tree lobster.

# Taylor's Checkerspot Butterfly

5

These butterflies live in North America. They drink from wildflowers. But roads and houses were built where the flowers grew. The butterfly was in big trouble.

Help came from an unusual place. A prison joined the effort to save this insect. Prisoners work with the Oregon Zoo. They help raise the caterpillars. When the butterflies are ready, the insects go back to the zoo. Then they are put back into the wild.

**Think About It**
What are some other groups that could work with zoos to save insects?

6

# Tusked Weta

The tusked weta is a cricket. It is the size of mouse. It lives on an island in New Zealand. But then rats arrived. They often killed this insect.

Scientists decided to help. They got rid of the rats. They also bred the crickets in **captivity**. They released the insect in new, rat-free places. Today, this giant bug is doing much better.

**Did You Know?**
The tusked weta is very smelly.
This makes it an easy target for rats.

## MORE TO EXPLORE

# WHERE IN THE WORLD?

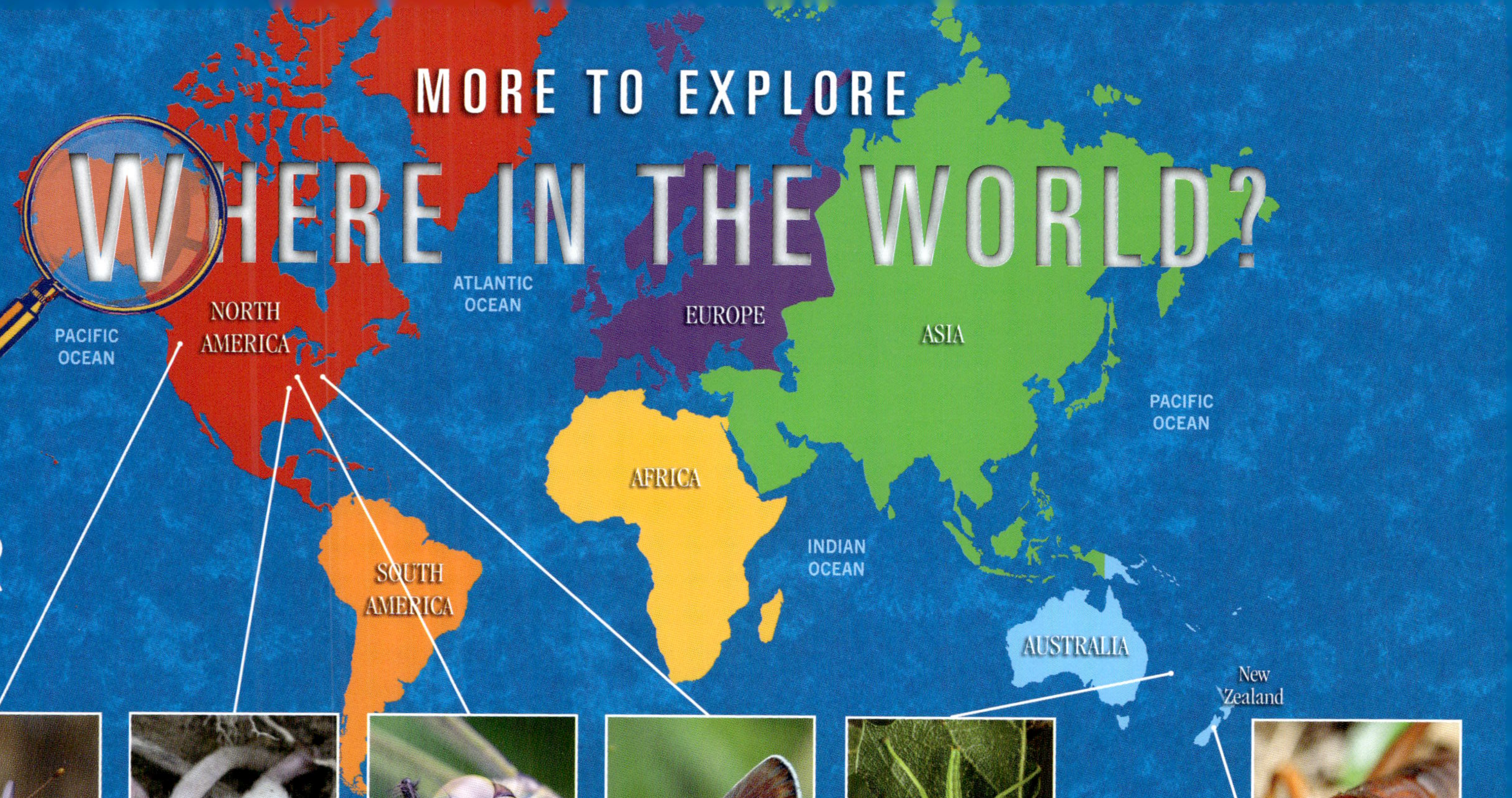

TAYLOR'S CHECKERSPOT BUTTERFLY
Northwestern United States and Canada

AMERICAN BURYING BEETLE
Midwestern and eastern United States

HINE'S EMERALD DRAGONFLY
Missouri and the Great Lakes area

KARNER BLUE BUTTERFLY
Northeastern United States

LORD HOWE ISLAND STICK INSECT
Ball's Pyramid, island near Australia

TUSKED WETA
New Zealand

MORE TO EXPLORE

# FANTASTIC FACTS

The American burying beetle is Rhode Island's state insect.

Female Hine's emerald dragonflies lay their eggs in water.

A Karner blue butterfly usually stays within 600 feet (183 meters) from where it hatches.

One Taylor checkerspot butterfly can lay up to 1,000 eggs.

Lord Howe Island stick insects cannot fly.

"Weta" means "god of ugly things" in the native Maori language.

MORE TO EXPLORE

# COOL COMPARISONS

How big are these insects?

**American Burying Beetle**
1–1.8 inches
(2.5–4.6 cm)

**Hine's Emerald Dragonfly**
2.5 inches
(6.4 cm)

**Karner Blue Butterfly**
1 inch
(2.5 cm)

**Lord Howe Island Stick Insect**
8 inches
(20 cm)

**Taylor Checkerspot Butterfly**
2.3 inches
(5.8 cm)

**Tusked Weta**
4 inches
(10 cm)

## MORE TO EXPLORE

# RESOURCES

### Glossary

**breed** (BREED) To give birth to or produce offspring.

**captivity** (kap-TIV-i-tee) The state of being kept in a place, such as a zoo.

**conservation** (kon-ser-VEY-shuhn) The act of keeping and protecting things from waste, loss, or destruction.

**habitat** (HAB-uh-tat) The place where a plant or animal grows or lives.

**larva** (LAR-vuh) *Plural* larvae; the wormlike form of an animal that hatches from an egg.

**legal** (LEE-gl) Of or having to do with law.

### Read More


Feldstein, Stephanie. *Save Pollinators*. Ann Arbor, MI: Cherry Lake Publishing, 2024.

Kelly, Christa. *Endangered Insects*. San Diego, CA: Brightpoint Press, 2024.


### Index




TOP RANK is published by Black Rabbit Books, P.O. Box 227, Mankato, MN, 56002. •



Top Rank is an imprint of Black Rabbit Books • Edited by Ana Brauer • Series designed by Danny Nanos • Book designed by Jason Knudson • Photographs © Alamy Stock Photo/MYN / Gil Wizen, 20; Dreamstime/Gutaper, 11, Paul Reeves, 18, Paul Sparks, 6–7; flickr/Jake Osborne, 19, 21, 23, Jill Utrup/USFWS, 21, 23; Getty Image/Matt Cardy, 13; Minden Pictures/John Abbott/NPL, 2; Shutterstock/Alexandra Lande, 12, Kanyshev Andrey, 9, 21, 23, LulaPar, cover, 1, Ramsey Samara, 17, 21, 23; SuperStock/Emanuele Biggi/Nature Picture Library, 14, 15, 21, 23, Michael Durham/Minden Pictures, 16; Wikimedia Commons/Bernard DUPONT, cover, 1, 10, Blocky1000, 4, Wayne National Forest, 5, 21, 23 • Printed in India

**Library of Congress Cataloging-in-Publication Data**: Names: Mattern, Joanne, 1963- author. | Title: Insects: back from the brink / by Joanne Mattern. | Description: Mankato, MN: Top Rank is an imprint of Black Rabbit Books, [2026] | Series: Back from the brink | Includes bibliographical references and index. | Audience: Ages 8–11 | Audience: Grades 2–3 | Identifiers: LCCN 2024043368 (print) | LCCN 2024043369 (ebook) | ISBN 9781644667989 (library binding) | ISBN 9781644668306 (paperback) | ISBN 9781644668627 (ebook) | Subjects: LCSH: Insects—Conservation—Juvenile literature. | Rare insects—Juvenile literature. | Rare invertebrates—Juvenile literature. | Classification: LCC QL467.8 .M382026 (print) | LCC QL467.8 (ebook) | DDC 333.9522—dc23/eng/20250108 | LC record available at https://lccn.loc.gov/2024043368 |LC ebook record available at https://lccn.loc.gov/2024043369